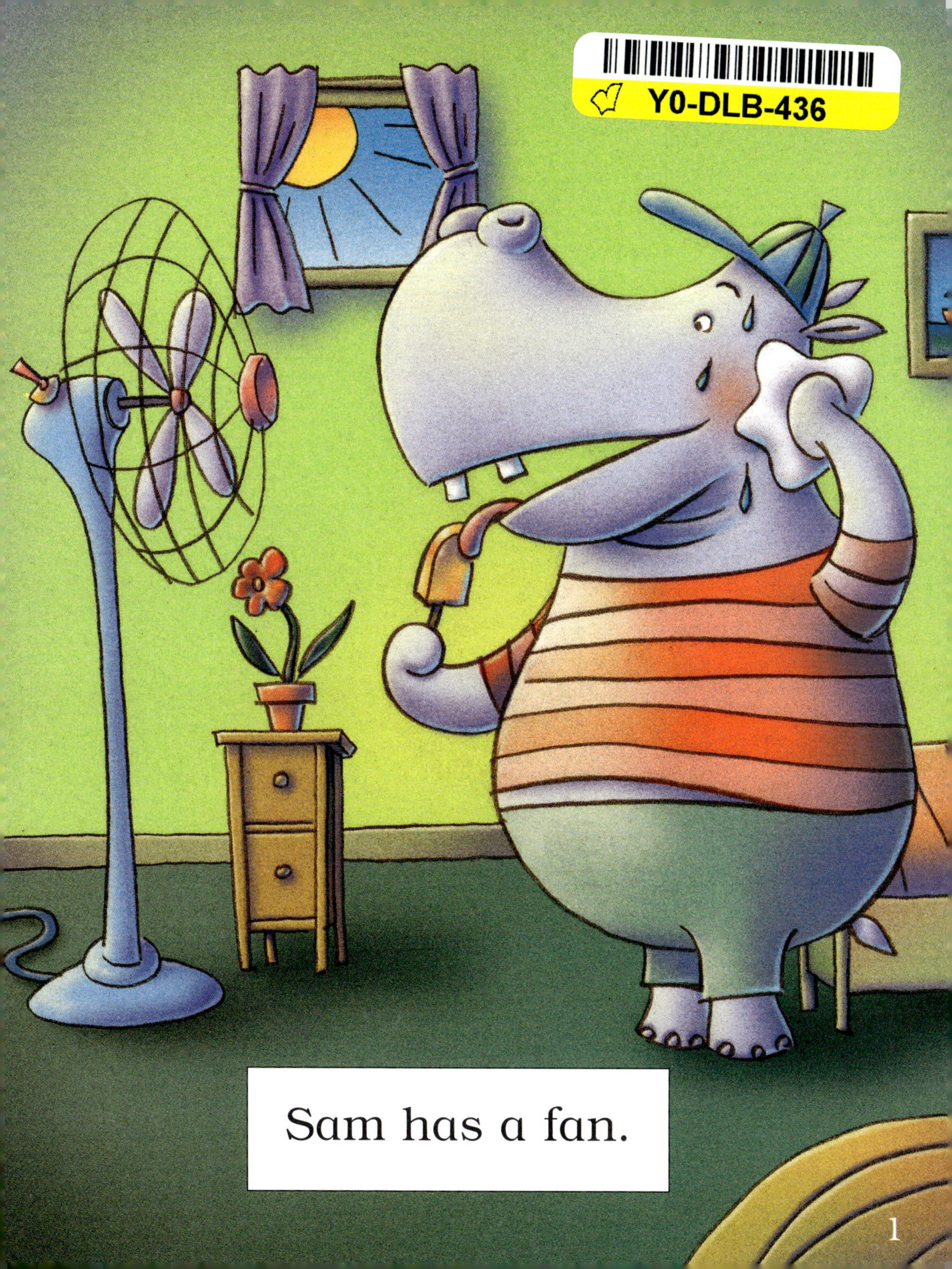

Sam has a fan.

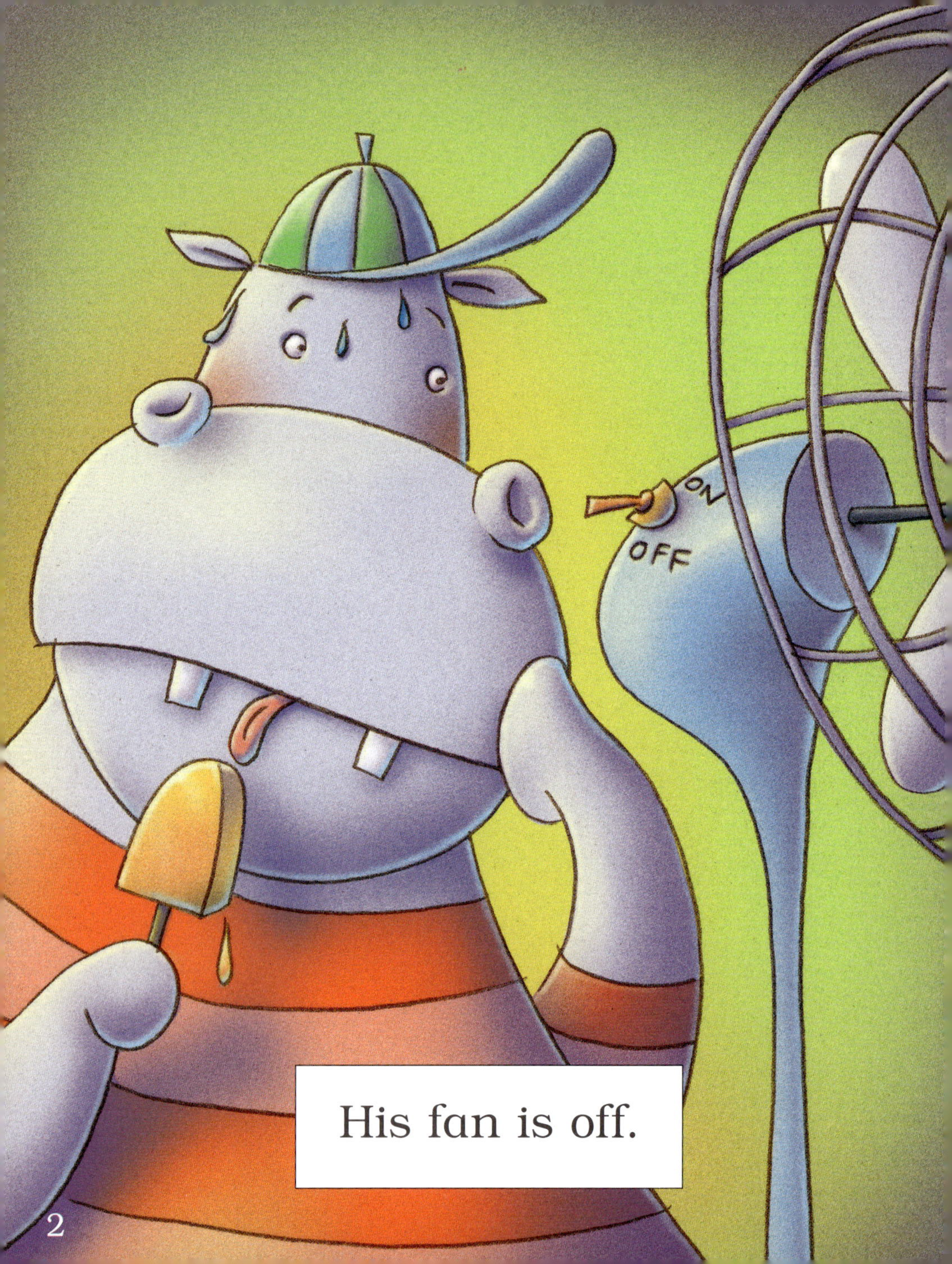

His fan is off.

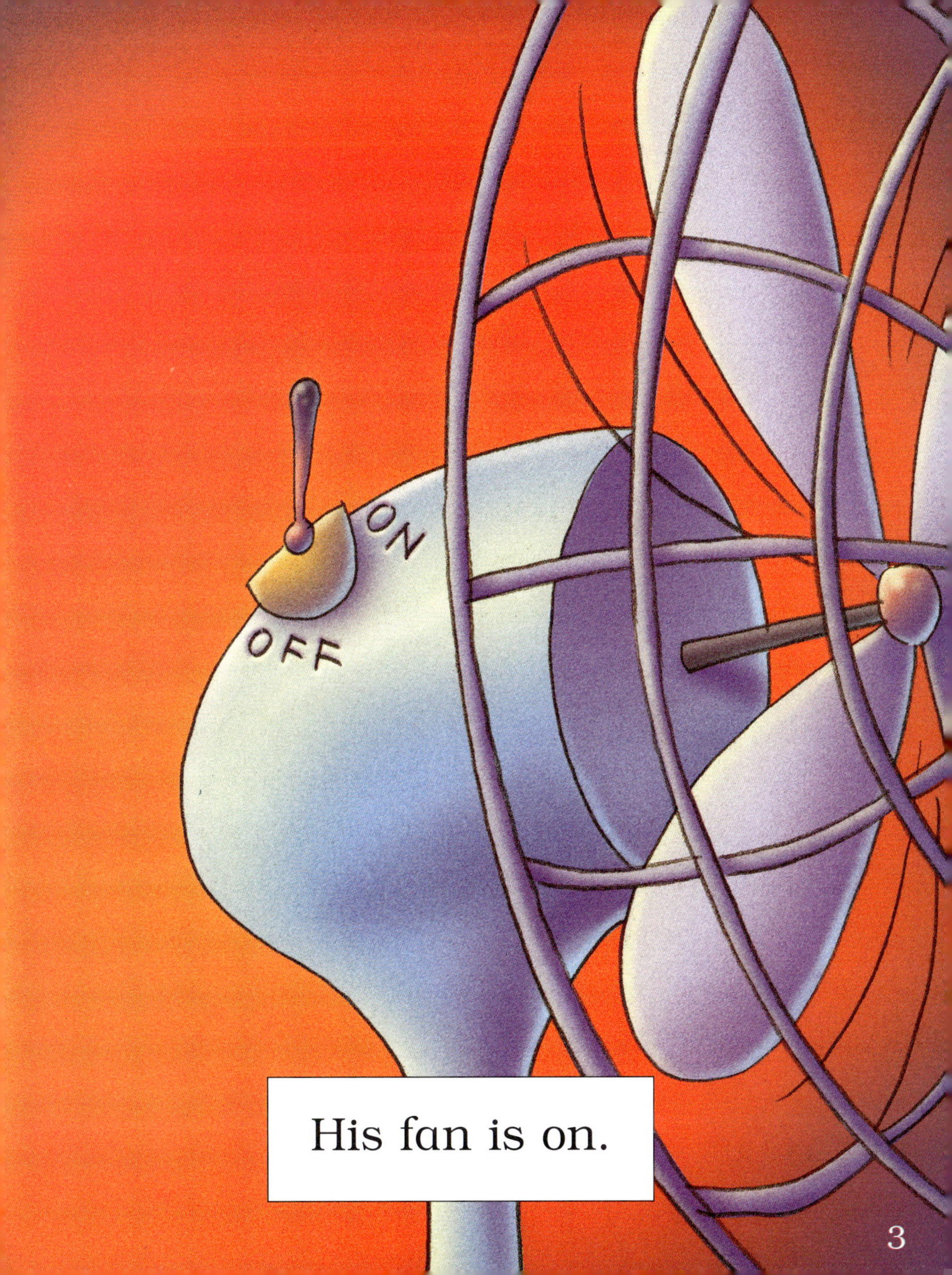

His fan is on.

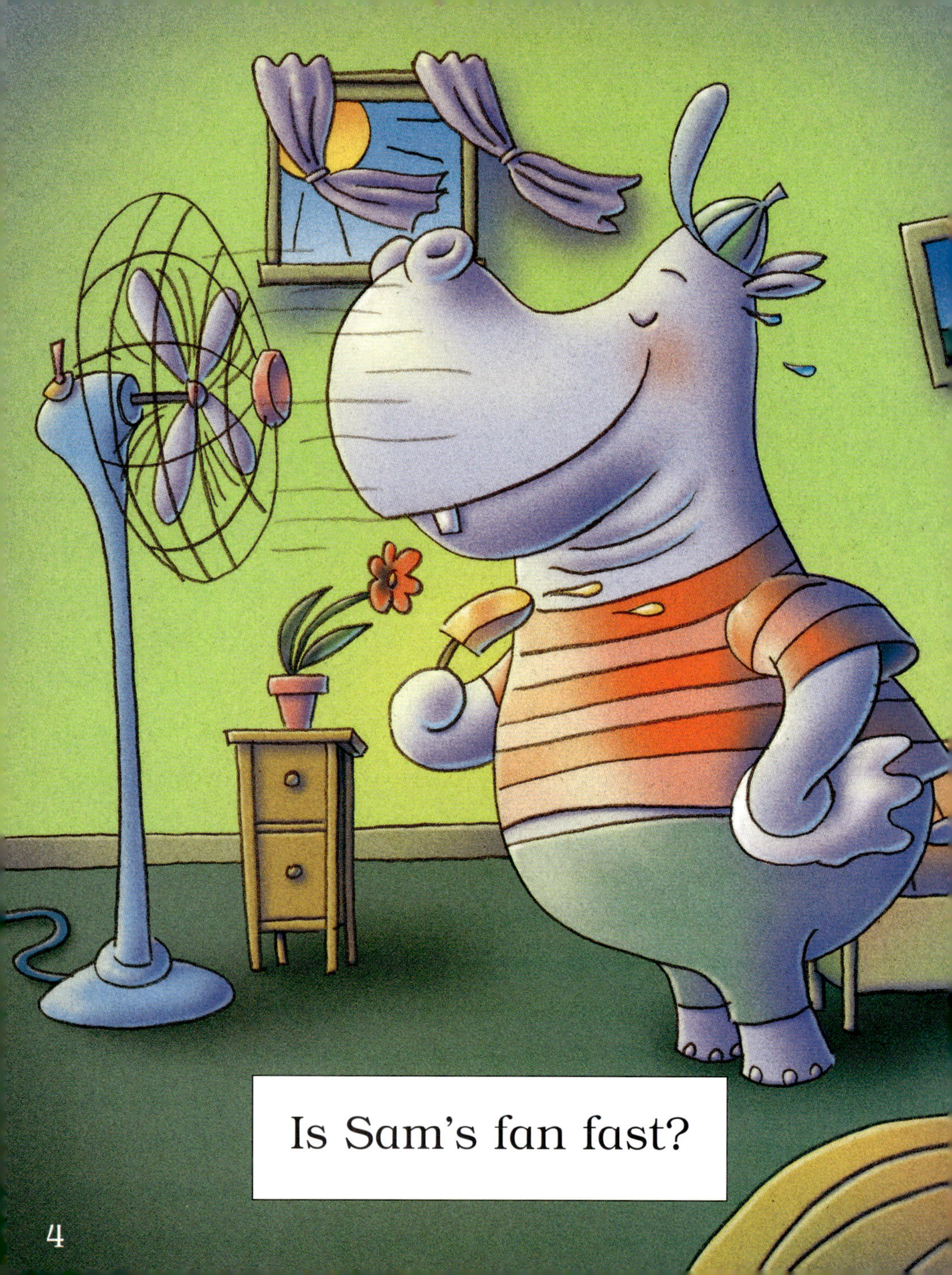

Is Sam's fan fast?

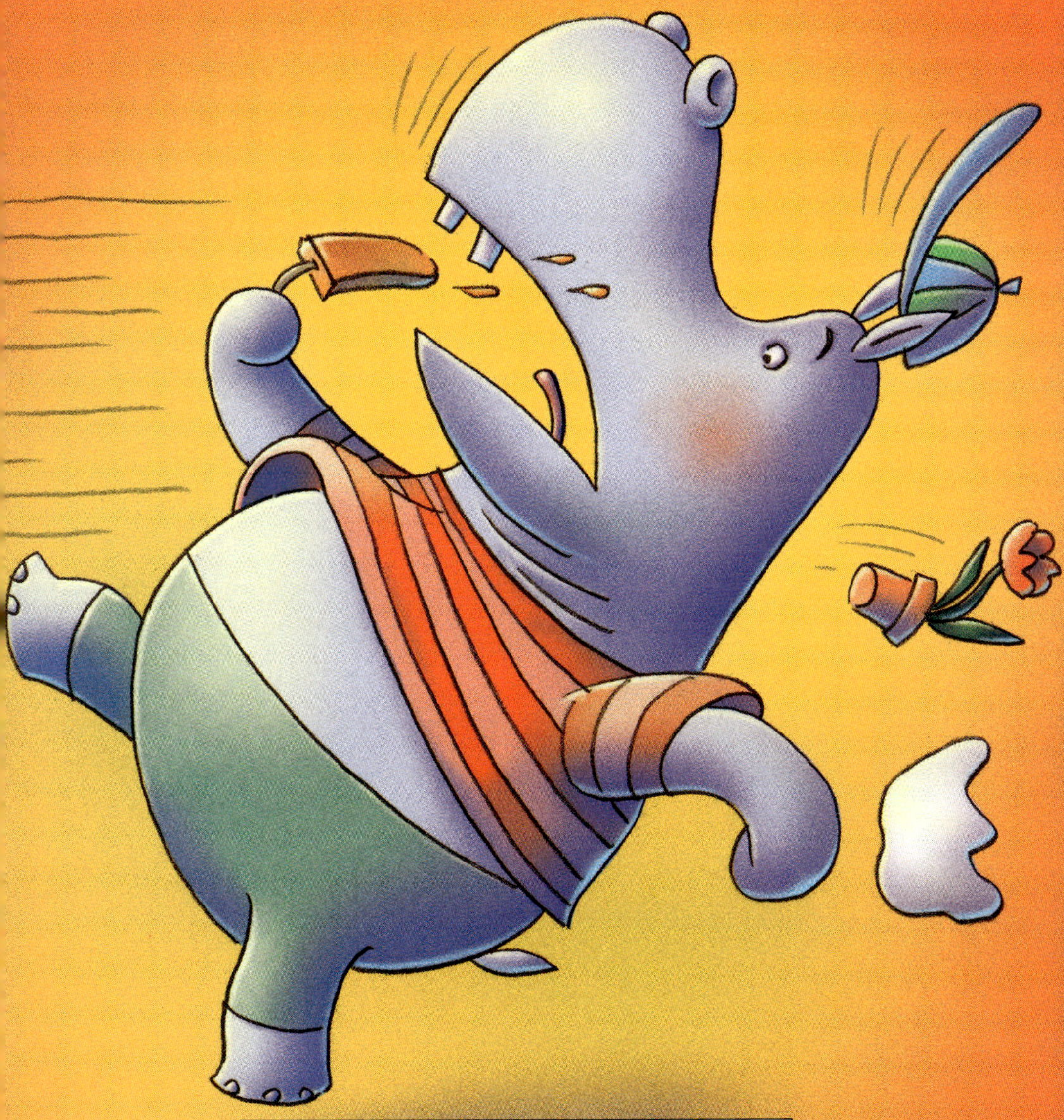

Stop the fan, Sam!

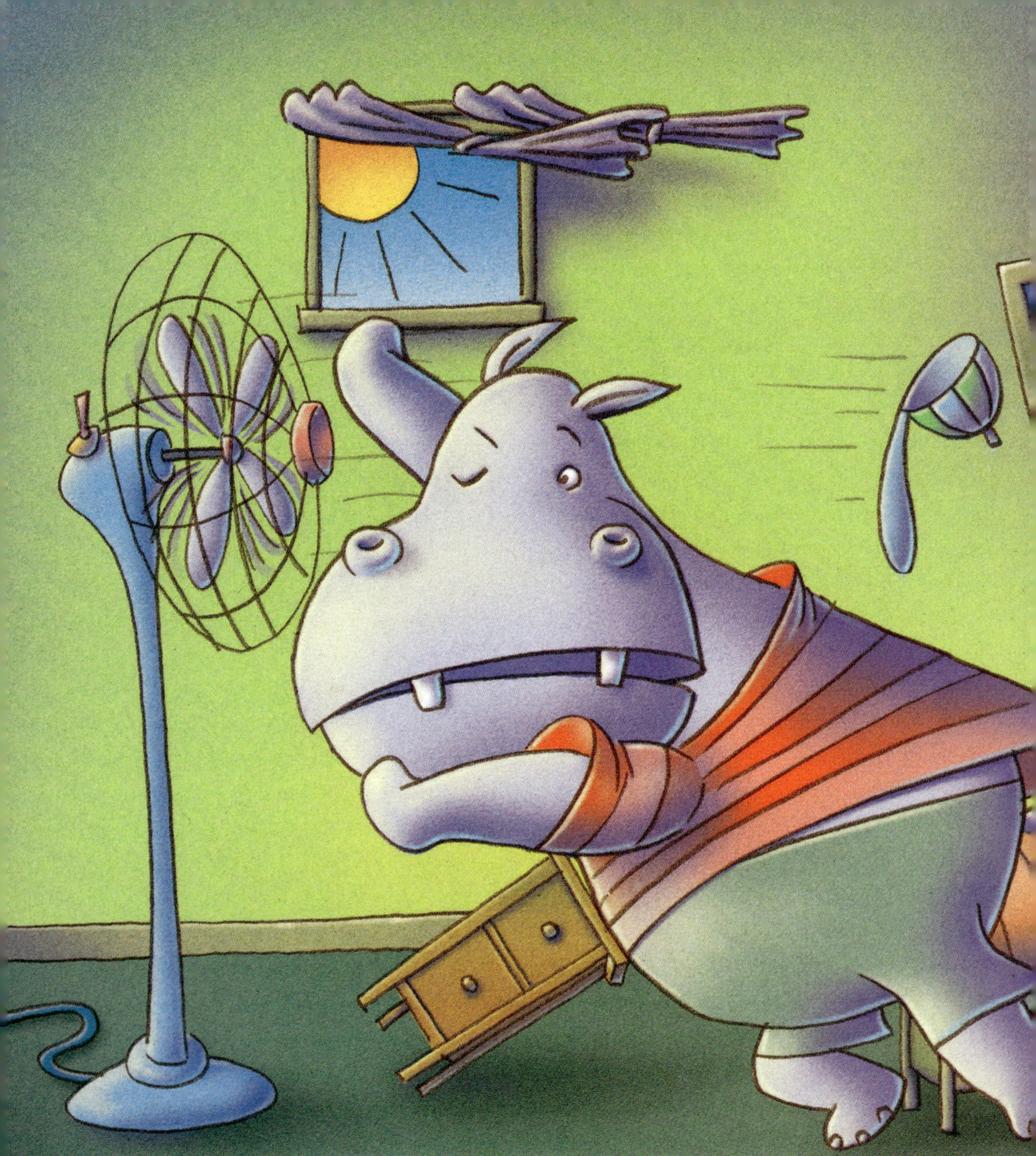

Stop the fan fast!

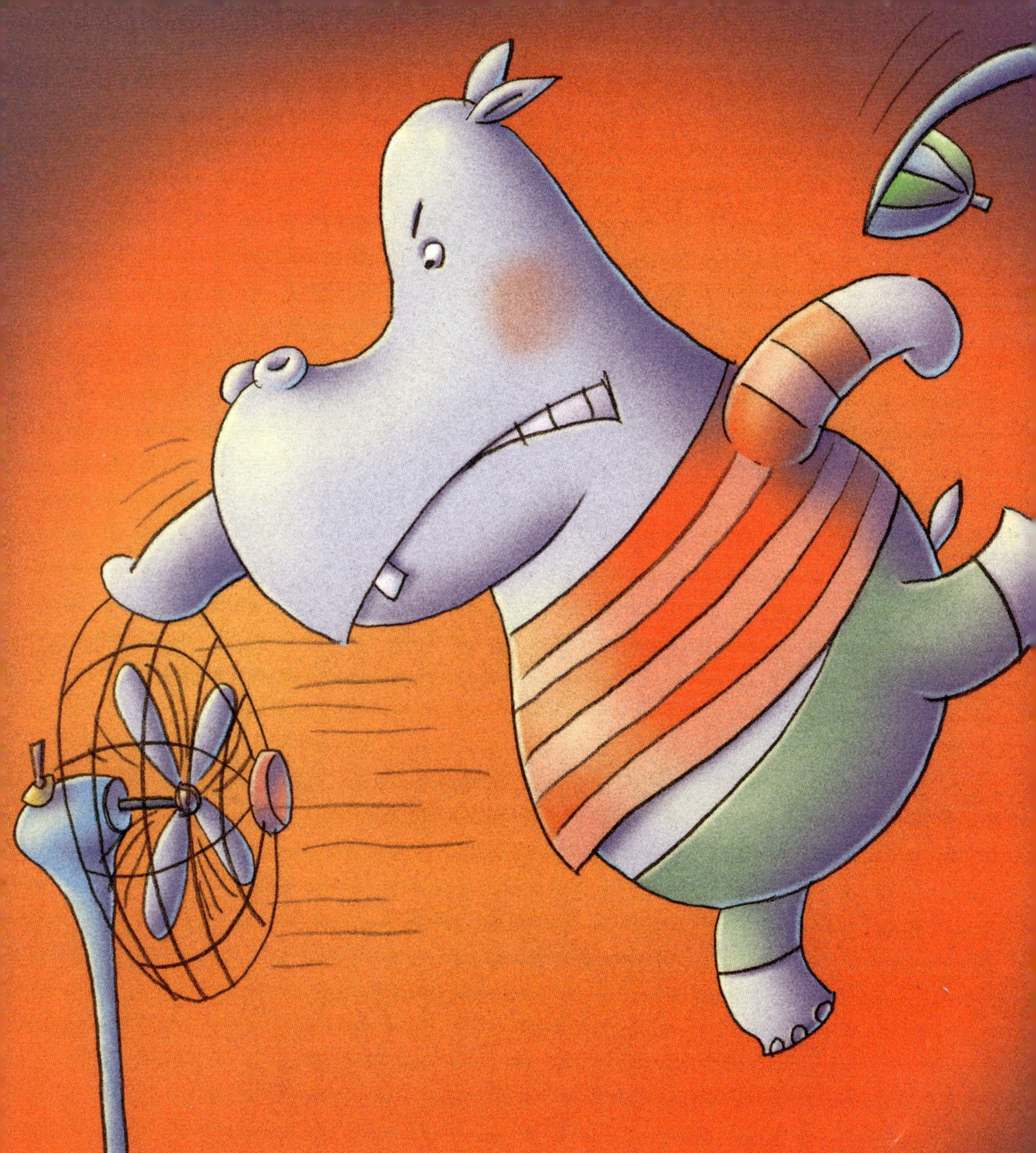

Sam hit the top.

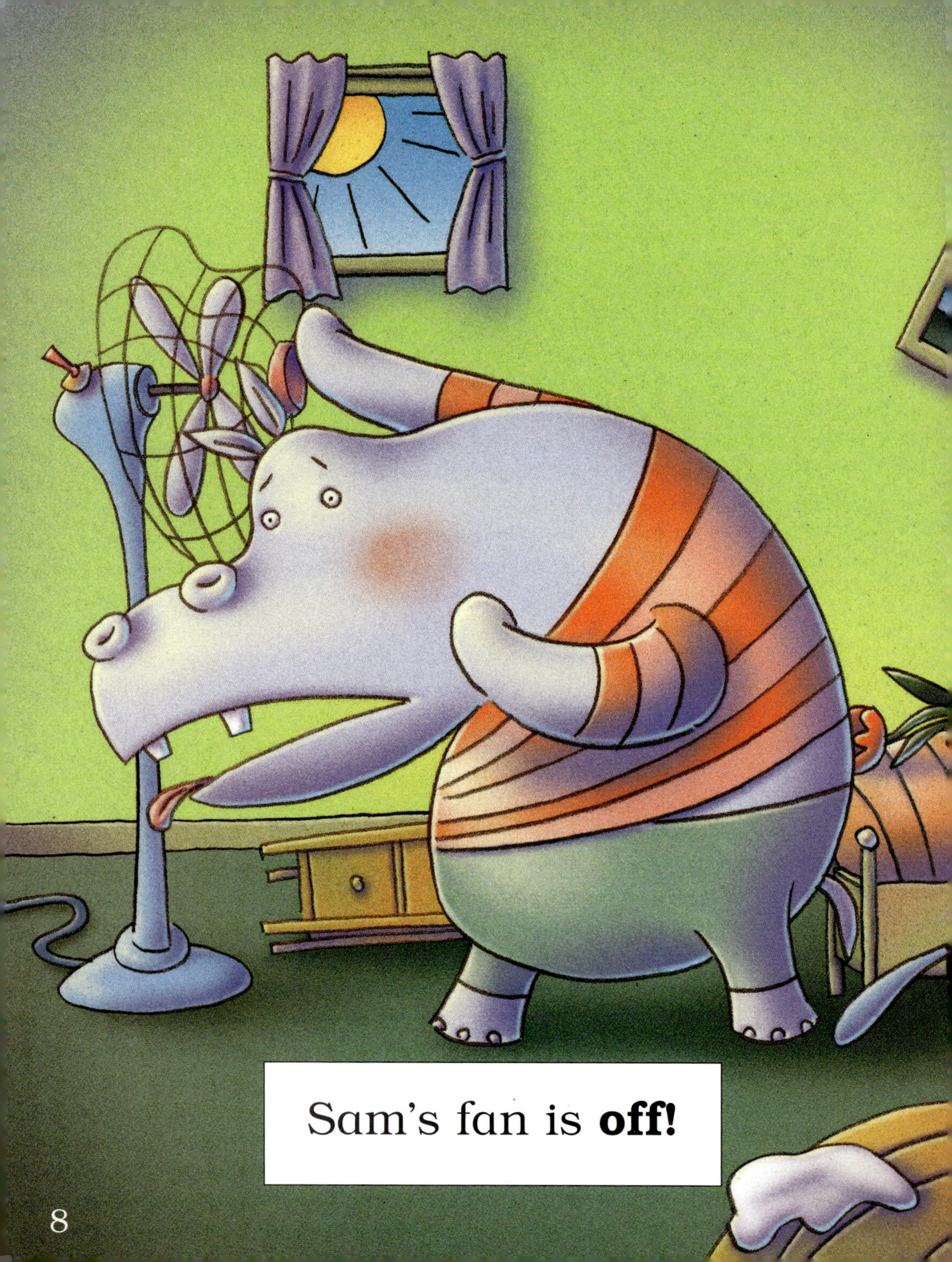

Sam's fan is **off!**